O9-ABI-607

SPACE MISSIONS™

The Gemini IV Mission

The First American Space Walk

Helen Zelon

The Rosen Publishing Group's
PowerKids Press™
New York

For Charlotte Zelon, whose brilliant inspiration endures with the stars

Published in 2002 by The Rosen Publishing Group, Inc.
29 East 21st Street, New York, NY 10010

First Edition

Book Design: Michael de Guzman

Project Editors: Jennifer Landau, Jason Moring, Jennifer Quasha

Photo Credits: p. 4 (constellation Gemini) © Roger Ressmeyer/CORBIS, (John Glenn and President Kennedy) © CORBIS; pp. 6, 7, 11 © Photri-Microstock; p. 8 (Vomit Comet) © NASA/Roger Ressmeyer/CORBIS, (astronauts at desert survival training) © Bettmann/CORBIS; pp. 12, 15 © Bettmann/CORBIS; p. 16 © NASA/Roger Ressmeyer/CORBIS; p. 19 (astronaut) © CORBIS, (gulf of California and Florida coastline) courtesy of NASA/JPL/California Institute of Technology; p. 20 (splashdown and recovery) © Photri-Microstock, (astronauts' return) Bettmann/CORBIS.

Zelon, Helen.
The Gemini IV Mission : the first American space walk / Helon Zelon.—1st ed.
p. cm.— (Space missions)
Includes index.
ISBN 0-8239-5771-3
1. Project Gemini (U.S.)—History—Juvenile literature. 2. Extravehicular activity (Manned space flight)—Juvenile literature.
[1. Project Gemini (U.S.) 2. Extravehicular activity (Manned space flight) 3. Manned space flight. 4. White, Edward Higgins, 1930–1967. 5. McDivitt, James Alton, 1929–] I. Title. II. Series.
TL89.8.U6 G97 2002
629.45'84—dc21

00-012001

Manufactured in the United States of America

Contents

This is a photograph of the constellation Gemini. ➔

➔ *In this photo, President John F. Kennedy and astronaut John Glenn stand next to the* Friendship 7 *space capsule.*

Project Gemini

Project Gemini was an important stepping-stone in America's path to the Moon. Gemini's name comes from the constellation, or group of stars, called Gemini.

In 1961, President John F. Kennedy announced that the United States would land a man on the Moon by 1970. Project Mercury was the first manned space program done by the United States. It proved that humans could live through space travel. At the end of the Mercury program, U.S. astronauts had spent fewer than 60 hours in space. That's only about half of the time it would take to fly to the Moon, one way!

NASA developed the Gemini program to continue the work Mercury began. Gemini let U.S. astronauts and scientists develop space survival skills. These skills would be important on future journeys to and from the Moon.

GEMINI 4
FIRST
SPACE WALK
McDIVITT
WHITE

Equipped to Explore

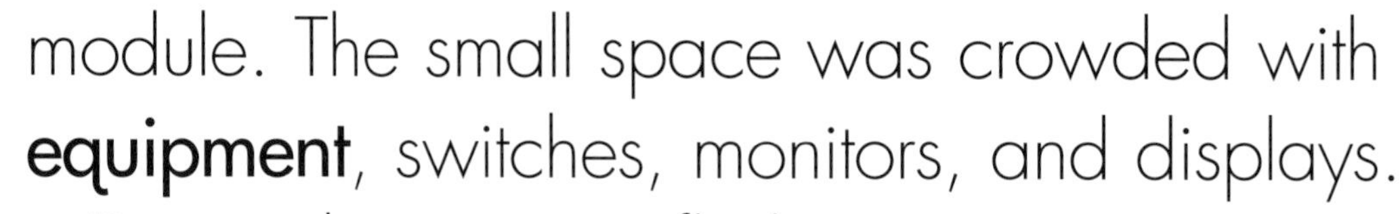

Each two-man Gemini space capsule was built in two sections. The astronauts lived and worked in the service module. The small space was crowded with **equipment**, switches, monitors, and displays.

During long spaceflights, astronauts needed oxygen to breathe, water to drink, and food to eat. All supplies were stored in the **adapter section**. The Gemini capsule also needed fuel for its rockets, and equipment to make electricity.

A strong rocket was needed to launch, or push, the space capsule into **orbit**. The Titan rockets used for Project Gemini were fast and

This picture shows the Gemini IV space capsule as it is launched, or pushed into space.

powerful. They were made in two sections, called **stages**. Each stage fell away from the space capsule once its fuel was used up.

Massive **friction** and heat are created during **re-entry**, the return to Earth's atmosphere. This heat made an intense fire that surrounded the spacecraft and the adapter section. The Gemini capsule and the astronauts were protected by a **heat shield** on the base of the spacecraft. The empty adapter section was designed to burst into flames on re-entry, leaving nothing behind.

This is a picture of astronauts Ed White and James McDivitt in the space capsule.

NASA
930

Basic Training

The Gemini IV astronauts were Edward H. White II and James A. McDivitt. Since childhood, Ed White had loved to fly. As an Air Force pilot, White flew training missions with astronauts. These practice missions let astronauts feel weightless, as they would in space. The astronauts got sick to their stomachs so often that the plane was nicknamed the Vomit Comet. White and McDivitt practiced for the mission for many hours on Earth. They went through desert survival training in case their spacecraft landed in the desert rather than in the ocean. It was important that the astronauts learn to depend on each other in emergencies. In space, no one would be able to help them. Their lives would depend on their training, their trust, and their friendship.

← Top left: *This is a photograph of the Vomit Comet, used for weightlessness training.* Bottom left: *This photograph shows White, McDivitt, and other astronauts training in the desert.*

Cool Tools

Project Gemini astronauts needed special equipment to help them live and work in space. Because there is no oxygen in space, Gemini astronauts wore bulky space suits to help them breathe. The space suits, helmets, and thick gloves were the astronauts' only protection against the extreme temperatures of space. Space walks required even more equipment. There was a backpack that held emergency oxygen in case the space suit failed, a ropelike object called an **umbilicus**, which kept the astronaut connected to the space capsule, and the newest tool, a "zip gun," that let an astronaut move around freely outside the spacecraft. The zip gun also could help an astronaut return to safety if the umbilicus broke.

This photograph shows an astronaut having his space suit checked for problems. ➔

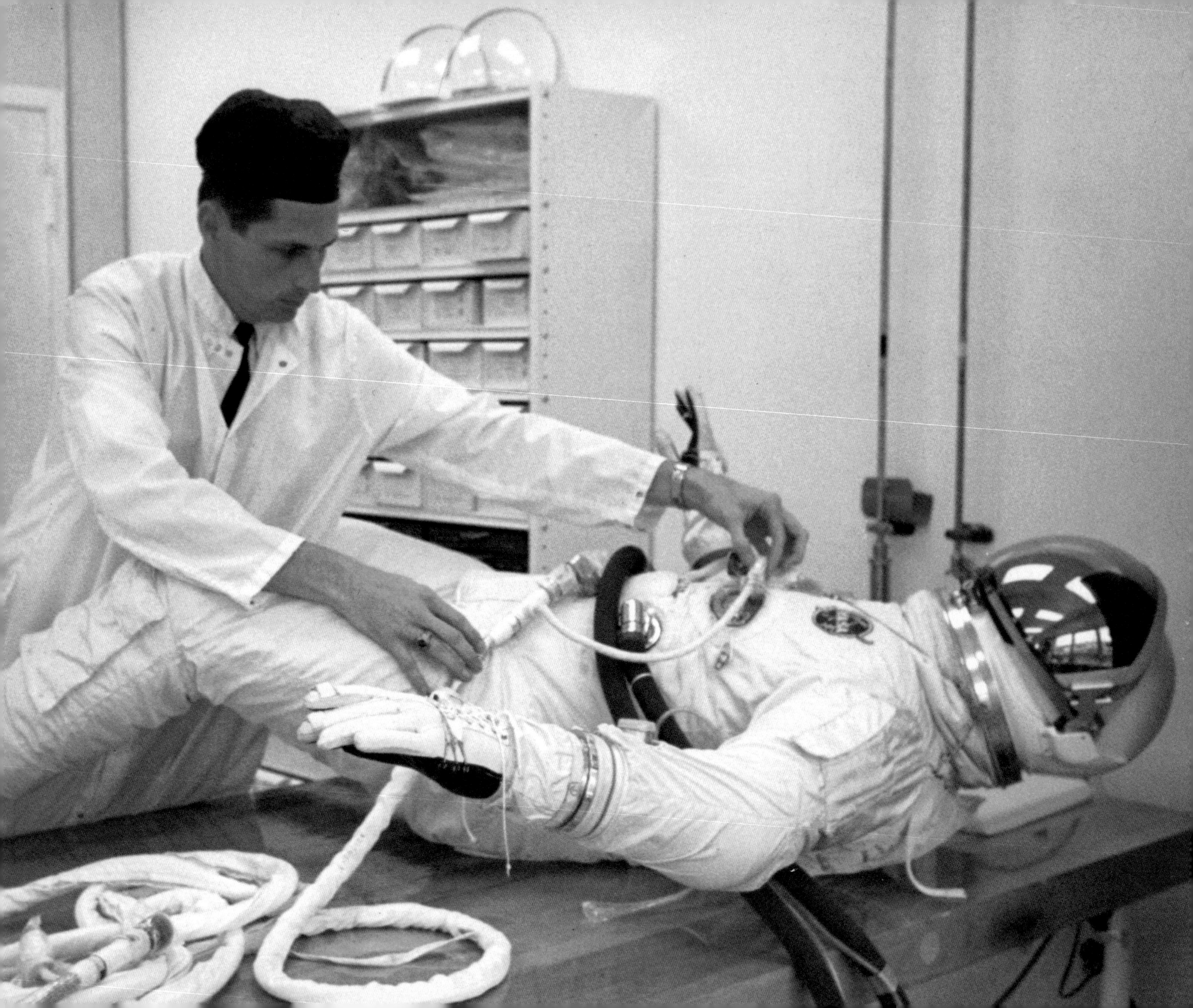

GEMINI 4
FIRST
SPACE WALK
McDIVITT
WHITE

Smooth Sailing

On June 3, 1965, by 7:00 A.M., Project Gemini astronauts Ed White and James McDivitt were dressed in their space suits. They rode the **gantry** elevator over 100 feet (31 meters) up and entered the space capsule.

Gemini IV was the first time American astronauts would try to meet, or rendezvous, with a released stage of the Titan rocket. Most exciting of all, Ed White would leave the Gemini space capsule for a space walk. He would take a stroll in space, powered by the brand-new zip gun. White, using the zip gun, would be the first man to actually move on his own in space. At 10:16 A.M., Gemini IV lifted off the **launchpad** and into a bright summer sky.

← This picture shows McDivitt and White entering the Gemini IV space capsule.

Playing Tag in Orbit

Gemini IV sailed high into space. Exploding bolts separated the rocket's **booster** from the space capsule. Jim McDivitt's job was to rendezvous with the booster and travel close to it. McDivitt tried to reach the booster without success. Every attempt used valuable fuel and the astronauts needed enough to return to Earth. Finally **Mission Control** told McDivitt and White to give up the rendezvous. It was time to move on to White's space walk. The Gemini IV astronauts and the engineers on the ground were surprised when the booster moved in a different way from what they had expected. Earth's gravity holds objects down, so they move in a straight line. In space, objects go forward AND they go up. There was still a lot to learn before human beings could go to the Moon.

This is a photo of astronauts White and McDivitt inside the space capsule. →

GEMINI 4
FIRST
SPACE WALK
McDIVITT
WHITE

First Steps in Space

Astronauts McDivitt and White prepared for White's space walk. They unpacked White's **chest pack**, which held oxygen for him to breathe. White put on his **thermal** gloves, because space was bitter cold. They unpacked a 25-foot-long (8-meter-long) umbilicus, White's lifeline to the Gemini capsule. Ed White opened the **hatch** overhead. He attached a camera to the outside of the space capsule. Before he left the capsule, White lowered a special visor on his helmet. The sun's light was blinding in space, but the visor allowed him to see. During the third Gemini orbit, Ed White left the capsule. White and Gemini IV were racing through the sky at 2,900 miles (4,667 km) a minute, but there was no noise! White was surprised by the quiet stillness. He watched the outlines of Earth beneath him.

← *This picture shows Ed White attached to the space capsule by a special cord called an umbilicus.*

GEMINI 4 FIRST SPACE WALK McDIVITT WHITE

"The Saddest Moment"

Ed White was the first American to walk in space. As he soared high above the United States, White saw the deep blue-green Pacific Ocean and the California beaches. He looked down on the Grand Canyon, the Rockies, the Mississippi River, and the Gulf of Mexico. Finally, when he saw the Florida seashore, he knew his space walk was nearly done. There was still time for a practical joke on Jim McDivitt, though. With bursts from the zip gun, White moved over to block McDivitt's window. White stamped his feet on the capsule's glass. "You smeared up my window, you dirty dog!" joked McDivitt from inside the capsule.

After 20 minutes outside the spacecraft, White's oxygen supply was running low. White sighed,

"It's the saddest moment of my life," as he headed back to the Gemini capsule.

Gemini IV continued in orbit, covering 1,609,700 miles (2,590,561 km). The astronauts took turns sleeping and working, planning for their return to Earth.

Top: *This is a photograph of White during his space walk.*
Bottom left: *This photograph shows the Florida seashore as seen from space.*
Bottom right: *This is a photograph of California beaches as seen from space.*

GEMINI 4
FIRST
SPACE WALK
McDIVITT
WHITE

Splashdown

After four days in orbit, it was time for Gemini IV's mission to end. McDivitt and White prepared for **splashdown** in the Atlantic Ocean. White and McDivitt fired their thrusters to adjust their approach to Earth. Soon after, the astronauts released the adapter section. The astronauts watched the adapter section burst into orange flames as it fell. Soon the Gemini capsule would plunge through the same fire, to safety. Back in Earth's atmosphere, the Gemini capsule's parachutes opened. The capsule fell into the Atlantic Ocean. The astronauts were rescued by helicopter and taken to an aircraft carrier, the *Wasp*. The sailors had rolled out a red carpet to greet them. Ed White felt so good that he danced a jig on the way to the crew's quarters.

Top: *This is a picture of the splashdown in the Atlantic Ocean.*
← Middle: *This photo shows astronauts leaving the capsule after splashdown.*
Bottom: *This is a photograph of White and McDivitt aboard the* Wasp.

GEMINI 4
FIRST
SPACE WALK
McDIVITT
WHITE

Bridge to the Moon

Gemini IV astronauts Ed White and Jim McDivitt were welcomed home as heroes. Together, White and McDivitt held the world record for the longest spaceflight. The two were in orbit for 98 hours. The president of the United States, Lyndon B. Johnson, honored each man with a special medal. President Johnson also invited White and McDivitt to represent the United States at the Paris Air Show. Pilots from all over the world would be there. The Gemini IV astronauts were greeted in Paris as American heroes and space pioneers. Their mission was a **milestone** in human spaceflight. White's pioneering space walk showed that astronauts could one day walk on the Moon's surface.

GEMINI 4 FIRST SPACE WALK McDIVITT WHITE

Glossary

adapter section (uh-DAP-ter SEK-shun) The space below a Gemini service module where supplies are stored.

booster (BOO-ster) The first section of a rocket that helps a space capsule head off into space.

chest pack (CHEST PAK) A self-contained unit worn by an astronaut that has emergency breathing equipment.

equipment (uh-KWIP-mint) All the supplies needed to do an activity.

friction (FRIK-shun) Energy, measured in heat, caused by two bodies of force in contact.

gantry (GAN-tree) A frame that supports a rocket on the launchpad.

hatch (HACH) A doorlike opening in a spacecraft.

heat shield (HEET SHEELD) A barrier that protects a spacecraft during reentry.

launchpad (LAWNCH-pad) The pad from which a spacecraft is sent off into the air.

milestone (MYL-stohn) A significant point in development, a marker.

Mission Control (MISH-shun kun-TROHL) A group of scientists and engineers who guide space travel from the ground.

orbit (OR-bit) The path one body makes around another, usually larger, body.

re-entry (ree-EN-tree) The return to Earth's atomspere from space.

stages (STAY-jez) Sections of a launch rocket.

splashdown (SPLASH-daun) The planned fall of a spacecraft into an ocean, where it is recovered.

thermal (THER-mul) Protecting against heat or cold.

umbilicus (uhm-BIL-ih-kus) A ropelike object that connects one thing to another.

Index

Web Sites

To learn more about Gemini IV, check out these Web sites:

http://spaceflight.nasa.gov

www.jsc.nasa.gov